ENERGY SECTOR STANDARD OF THE PEOPLE'S REPUBLIC OF CHINA
中华人民共和国能源行业标准

Guide for Acceptance of Archives for Hydropower Projects

水电工程项目档案验收工作导则

NB/T 10076-2018

Chief Development Department: China Renewable Energy Engineering Institute
Approval Department: National Energy Administration of the People's Republic of China
Implementation Date: March 1, 2019

China Water & Power Press
中国水利水电出版社
Beijing 2024

All rights reserved. No part of this publication may be reproduced, stored in a retrieval system, or transmitted in any form or by any means—electronic, mechanical, photocopying, recording or otherwise, without prior written permission of the publisher.

图书在版编目（CIP）数据

水电工程项目档案验收工作导则：NB/T 10076-2018 = Guide for Acceptance of Archives for Hydropower Projects(NB/T 10076-2018)：英文 / 国家能源局发布. 北京：中国水利水电出版社, 2024. 10. -- ISBN 978-7-5226-2769-4

I. G275.3-65

中国国家版本馆CIP数据核字第2024LW7490号

ENERGY SECTOR STANDARD
OF THE PEOPLE'S REPUBLIC OF CHINA
中华人民共和国能源行业标准

Guide for Acceptance of Archives for Hydropower Projects
水电工程项目档案验收工作导则
NB/T 10076-2018
（英文版）

Issued by National Energy Administration of the People's Republic of China
国家能源局　发布
Translation organized by China Renewable Energy Engineering Institute
水电水利规划设计总院　组织翻译
Published by China Water & Power Press
中国水利水电出版社　出版发行
　　Tel: (+ 86 10) 68545888　68545874
　　sales@mwr.gov.cn
　　Account name: China Water & Power Press
　　Address: No.1, Yuyuantan Nanlu, Haidian District, Beijing 100038, China
　　http: //www.waterpub.com.cn
中国水利水电出版社微机排版中心　排版
北京中献拓方科技发展有限公司　印刷
184mm×260mm　16开本　2.75印张　87千字
2024年10月第1版　2024年10月第1次印刷
Price（定价）：￥440.00

Introduction

This English version is one of China's energy sector standard series in English. Its translation was organized by China Renewable Energy Engineering Institute authorized by National Energy Administration of the People's Republic of China in compliance with relevant procedures and stipulations. This English version was issued by National Energy Administration of the People's Republic of China in Announcement [2021] No. 6 dated December 22, 2021.

This version was translated from the Chinese Standard NB/T 10076-2018, *Guide for Acceptance of Archives for Hydropower Projects*, published by China Water & Power Press. The copyright is reserved by National Energy Administration of the People's Republic of China. In the event of any discrepancy in the implementation, the Chinese version shall prevail.

Many thanks go to the staff from the relevant standard development organizations and those who have provided generous assistance in the translation and review process.

For further improvement of the English version, any opinions and suggestions are welcome and should be addressed to:

China Renewable Energy Engineering Institute
No. 2 Beixiaojie, Liupukang, Xicheng District, Beijing 100120, China
Website: www.creei.cn

Translating organizations:

HYDROCHINA CORPORATION LIMITED

China Renewable Energy Engineering Institute

Translating staff:

YANG Jing	LIU Qianhua	WU Hehe	SONG Dianhai
CHEN Zhongxing	GUAN Chunxi	LI Yanzhi	ZHOU Ming
CHE Zhenying	ZHU Lin		

Review panel members:

LIU Xiaofen	POWERCHINA Zhongnan Engineering Corporation Limited
QIE Chunsheng	Senior English Translator
QI Wen	POWERCHINA Beijing Engineering Corporation Limited

JIN Feng	Tsinghua University
YAN Wenjun	Army Academy of Armored Forces, PLA
CHEN Jiansu	HYDROCHINA CORPORATION LIMITED
QIAO Peng	POWERCHINA Northwest Engineering Corporation Limited
HOU Hongying	China Renewable Energy Engineering Institute
LIU Jianlin	POWERCHINA Chengdu Engineering Corporation Limited
ZHANG Qian	POWERCHINA Guiyang Engineering Corporation Limited

National Energy Administration of the People's Republic of China

翻译出版说明

本译本为国家能源局委托水电水利规划设计总院按照有关程序和规定，统一组织翻译的能源行业标准英文版系列译本之一。2021 年 12 月 22 日，国家能源局以 2021 年第 6 号公告予以公布。

本译本是根据中国水利水电出版社出版的《水电工程项目档案验收工作导则》NB/T 10076—2018 翻译的，著作权归国家能源局所有。在使用过程中，如出现异议，以中文版为准。

本译本在翻译和审核过程中，本标准编制单位及编制组有关成员给予了积极协助。

为不断提高本译本的质量，欢迎使用者提出意见和建议，并反馈给水电水利规划设计总院。

地址：北京市西城区六铺炕北小街 2 号
邮编：100120
网址：www.creei.cn

本译本翻译单位：中国水电工程顾问集团有限公司
　　　　　　　　水电水利规划设计总院

本译本翻译人员：杨　静　刘倩华　吴鹤鹤　宋殿海
　　　　　　　　陈中兴　关春溪　李彦质　周　明
　　　　　　　　车振英　朱　琳

本译本审核人员：
　　刘小芬　中国电建集团中南勘测设计研究院有限公司
　　郄春生　英语高级翻译
　　齐　文　中国电建集团北京勘测设计研究院有限公司
　　金　峰　清华大学
　　闫文军　中国人民解放军陆军装甲兵学院基础部
　　陈建苏　中国水电工程顾问集团有限公司
　　乔　鹏　中国电建集团西北勘测设计研究院有限公司
　　侯红英　水电水利规划设计总院
　　刘建琳　中国电建集团成都勘测设计研究院有限公司
　　张　倩　中国电建集团贵阳勘测设计研究院有限公司

国家能源局

Announcement of National Energy Administration of the People's Republic of China [2018] No. 12

According to the requirements of Document GNJKJ [2009] No. 52, "Notice on Releasing the Energy Sector Standardization Administration Regulations (*tentative*) and detailed implementation rules issued by National Energy Administration of the People's Republic of China", 204 energy sector standards such as *Coal Mine Air-Cooling Adjustable-Speed Magnetic Coupling*, including 54 energy standards (NB), 8 petrochemical standards (NB/SH), and 142 petroleum standards (SY), are issued by National Energy Administration of the People's Republic of China after due review and approval.

Attachment: Directory of Sector Standards

National Energy Administration of the People's Republic of China

October 29, 2018

Attachment:

Directory of Sector Standards

Serial number	Standard No.	Title	Replaced standard No.	Adopted international standard No.	Approval date	Implementation date
...						
31	NB/T 10076-2018	Guide for Acceptance of Archives for Hydropower Projects			2018-10-29	2019-03-01
...						

Foreword

According to the requirements of Document GNKJ [2015] No. 12 issued by National Energy Administration of the People's Republic of China, "Notice on Releasing the Development and Revision Plan of the Second Batch of Energy Sector Standards in 2014", and after extensive investigation and research, summarization of practical experience, and wide solicitation of opinions, the drafting group has prepared this guide.

The main technical contents of this guide include: acceptance organization, acceptance preconditions, acceptance content, and acceptance procedures.

National Energy Administration of the People's Republic of China is in charge of the administration of this guide. China Renewable Energy Engineering Institute has proposed this guide and is responsible for its routine management. Energy Sector Standardization Technical Committee on Hydropower Investigation and Design is responsible for the explanation of specific technical contents. Comments and suggestions in the implementation of this guide should be addressed to:

China Renewable Energy Engineering Institute
No. 2 Beixiaojie, Liupukang, Xicheng District, Beijing 100120, China

Chief development organizations:

China Renewable Energy Engineering Institute

HYDROCHINA CORPORATION LIMITED

Huanghe Hydropower Development Co., Ltd.

Participating development organizations:

YALONG RIVER HYDROPOWER DEVELOPMENT COMPANY LTD.

POWERCHINA Zhongnan Engineering Corporation Limited

Chief drafting staff:

WU Hehe	NI Ping	SU Xiaojun	ZHANG Dingrong
MI Biao	FAN Jianzhen	DENG Xuehui	WANG Side
ZHANG Jie	WANG Weibing	YAO Liping	ZHOU Xianrui
WANG Lingling	ZHU Chunmei	LI Hua	

Review panel members:

| PENG Caide | WANG Yanmin | YU Hui | WANG Hongmin |

LIN Hong	ZHAO Xinyu	ZHANG Jian	XU Jun
WANG Lei	HU Xinghua	PAN Bin	WANG Liqun
NIU Hongli	HOU Hongying	MA Lin	RU Xin
DU Gang			

Contents

1 General Provisions ································· 1
2 Terms ······································· 2
3 Acceptance Organization ··························· 3
4 Acceptance Preconditions ·························· 4
5 Acceptance Content ······························ 6
6 Acceptance Procedures ··························· 7
  6.1 General Requirements ·························· 7
  6.2 Acceptance Preparation ························ 7
  6.3 Site Acceptance ···························· 8
  6.4 Implementation of Acceptance Opinions ················ 9
Appendix A Archives Acceptance Workflow for Hydropower Projects ································ 11
Appendix B Main Contents of Self-Inspection Report by Project Owner ······························ 12
Appendix C Main Contents of Self-Inspection Report by Designer ································ 13
Appendix D Main Contents of Self-Inspection Report by Supervisor ······························· 14
Appendix E Main Contents of Self-Inspection Report by Construction Contractor ······················ 15
Appendix F Main Contents of Self-Inspection Report by Units for Testing, Inspection and Safety Monitoring ··· 16
Appendix G Main Contents of Archives Self-Inspection Report by EPC Contractor ······················ 17
Appendix H Self-Evaluation Form for Completion Archives Acceptance of Hydropower Projects ············· 18
Appendix J Application Form for Archives Acceptance of Hydropower Projects ······················· 29
Explanation of Wording in This Guide ····················· 30
List of Quoted Standards ··························· 31

1 General Provisions

1.0.1 This guide is formulated with a view to providing guidance on archives acceptance and management for hydropower projects.

1.0.2 This guide is applicable to the archives acceptance for the construction, renovation and extension of large- and medium-sized hydropower projects, but not applicable to the archives acceptance for resettlement works which is administrated by the government.

1.0.3 The archives acceptance for hydropower projects is an important part of the project completion acceptance. Project completion acceptance shall not be carried out unless the project archives acceptance has been conducted and rated as acceptable.

1.0.4 The archives acceptance for hydropower projects is divided into staged archives acceptance and completion archives acceptance. Completion archives acceptance shall be conducted for all hydropower projects. An archives pre-acceptance may be organized before the completion archives acceptance, depending on the project scale, construction period and complexity. Archives acceptance should be conducted for Large-size (1) hydropower projects at the river closure and initial impoundment stages, for Large-size (2) hydropower projects at the river closure stage, and for the pumped storage power projects at the initial impoundment stage.

1.0.5 In addition to this guide, the archives acceptance for hydropower projects shall comply with other current relevant standards of China.

2 Terms

2.0.1 records of project

documents of a project in the form of texts, graphs, audios, and videos generated throughout the project development from initiation, bidding, investigation, design, equipment and material procurement, construction, supervision, testing, and commissioning to completion acceptance

2.0.2 archival arrangement

process of systematic sequencing of records of project including classification, grouping, arranging, numbering and cataloging in accordance with relevant rules

2.0.3 archives of project

records of a project that have been appraised, arranged and filed

2.0.4 file

filing unit consisting of a group of related documents

3 Acceptance Organization

3.0.1 The organizer for the completion archives acceptance of a hydropower project shall be designated in accordance with the relevant national laws, regulations and rules.

3.0.2 The staged archives acceptance for a hydropower project may be organized by the project competent organization.

3.0.3 The project archives acceptance team shall be established according to the following requirements:

1. For the project archives acceptance organized by the national archives administration, the acceptance team shall consist of persons from the national archives administration, the provincial archives administration, and the project owner's superior.

2. For the project archives acceptance organized by the provincial archives administration, the acceptance team shall consist of persons from the provincial archives administration and the project owner's superior.

3. For the project archives acceptance organized by the project owner's superior, the acceptance team shall consist of persons from the project owner's superior and the provincial archives administration.

4. The acceptance team shall consist of archivists and hydropower technical experts. The number of the project archives acceptance team members should be odd and not less than 7, and the team shall be chaired by the archives acceptance organizer.

4 Acceptance Preconditions

4.0.1 The following preconditions shall be satisfied before the archives acceptance at the river closure stage:

1. The collection, arrangement and filing of records of project at early stages have been completed.
2. The filing and transfer of records of the completed and accepted contract works have been completed.
3. The collection and arrangement of records of the river closure works have been completed.
4. The project owner has completed the self-inspection of archives at the river closure stage and has prepared the self-inspection report.

4.0.2 The following preconditions shall be satisfied before the archives acceptance at the initial impoundment stage:

1. The filing and transfer of records of the completed and accepted contract works have been completed.
2. The collection and arrangement of records of the initial impoundment works have been completed.
3. The parties concerned, including project owner, supervisor, main construction contractors, and units for testing, inspection and safety monitoring, have completed their respective self-inspection on the archives at the initial impoundment stage and have prepared their self-inspection reports.

4.0.3 The following preconditions shall be satisfied before the completion archives acceptance:

1. The hydropower complex has been completed in line with the approved design scale and standard, and put into production and use. A small number of outstanding works are allowed but they shall not affect the safe and normal operation of the project.
2. The hydropower project has passed the commissioning test, and each generating unit has been running normally for more than 2000 h, which includes the standby time for the pumped storage power station.
3. All works, except some special works, have been operating normally and meet the design and functional requirements.
4. The collection, arrangement, filing and transfer of records of the project

have been completed in accordance with the current standards of China GB/T 11822, *General Requirements for the File Formation of Scientific and Technological Archives*; DA/T 28, *Specification for Archival Management of Construction Projects*; DL/T 1396, *Specification of the Documents Collection and Archives Arrangement for Hydropower Construction Project*s; and NB/T 35083, *Specification for Drafting As-built Drawing Documents of Hydropower Projects.*

5   The parties concerned, including project owner, designer, supervisor, main construction contractors and units for testing, inspection and safety monitoring, have completed their respective archives self-inspection and have prepared their self-inspection reports.

6   The project owner has completed the archives self-evaluation, which is deemed acceptable.

5 Acceptance Content

5.0.1 The archives acceptance at the river closure stage shall cover the following:

1. Project archives organizational system and staffing.
2. Archives management system, including the records classification and filing scope.
3. Filing of records of project at early stages.
4. Filing of records of the completed and accepted contract works.
5. Generation, collection and arrangement of records relating to the river closure.
6. Archives custody and security.

5.0.2 The archives acceptance at the initial impoundment stage shall cover the following:

1. Operation of project archives organizational system and execution of management system.
2. Filing of records of the completed and accepted contract works.
3. Generation, collection and arrangement of records relating to the initial impoundment.
4. Archives security and informatization.

5.0.3 The completion archives acceptance shall cover the following:

1. Archives support system and its execution.
2. Compilation of as-built drawings and its quality.
3. Archives completeness, accuracy and systematicness.
4. Archives security, accessibility and informatization.
5. Rectification of the problems found in the previous archives acceptances and its feedback.

6 Acceptance Procedures

6.1 General Requirements

6.1.1 The archives acceptance procedures for hydropower projects shall cover acceptance preparation, site acceptance, and implementation of acceptance opinions. The archives acceptance workflow for hydropower project should be in accordance with Appendix A of this guide.

6.1.2 Parties concerned shall ensure the authenticity and accuracy of the records submitted for acceptance.

6.1.3 Archives acceptance shall be concluded as acceptable or unacceptable.

6.2 Acceptance Preparation

6.2.1 Prior to the archives acceptance at the river closure stage, the project owner shall carry out the archives self-inspection and prepare the self-inspection report.

6.2.2 Prior to the archives acceptance at the initial impoundment stage, the project owner shall organize the parties concerned, including supervisor, main construction contractors, and units for testing, inspection and safety monitoring to carry out archives self-inspection and prepare their self-inspection reports, which should be in accordance with Appendixes B, C, D, E, and F of this guide.

6.2.3 Prior to the completion archives acceptance, the project owner shall organize the parties concerned, including designer, supervisor, main construction contractors, and units for testing, inspection and safety monitoring, to carry out archives self-inspection and prepare their self-inspection reports.

6.2.4 For EPC projects, the prime contractor is responsible for preparing the archives self-inspection report for designer and main construction contractors, which should be in accordance with Appendix G of this guide.

6.2.5 Prior to the completion archives acceptance, the project owner shall, on the basis of self-inspection, conduct the archives self-evaluation in accordance with Appendix H of this guide, and the self-evaluation conclusion shall be attached to the archives self-inspection report. All the dominant items shall be compliant, and the compliance rate of total items shall reach 80 % or above.

6.2.6 The project owner shall provide supporting documents for archives acceptance, including:

1 Project archives management rules.

2 Training records and instructions on project archival work.

    3   Project breakdown table, list of bidding works, list of contracts, and list of equipment.

    4   Project archives classification and filing scope.

    5   Retrieval aids.

    6   Archives accessibility.

    7   Opinions of staged archives acceptance and pre-acceptance, and rectifications.

6.2.7 After the archives self-inspection passes, the project owner shall send a letter of application for archives acceptance to the archives acceptance organizer, which shall be attached with an application form and the self-inspection report. The archives acceptance form should be in accordance with Appendix J of this guide.

6.2.8 After receiving the application for archives acceptance, the acceptance organizer shall conduct a preliminary examination.

6.2.9 After the preliminary examination passes, the acceptance organizer should set up an acceptance team in accordance with Article 3.0.3 of this guide to carry out the archives acceptance.

6.2.10 The project owner shall organize the parties concerned to participate in the site acceptance.

6.3   Site Acceptance

6.3.1 The archives site acceptance for a hydropower project should include the preparatory meeting of the acceptance team, kick-off meeting, site visit, archives inspection, internal meeting of the acceptance team, and wrap-up meeting.

6.3.2 The preparatory meeting of the acceptance team shall be presided over by the team leader and attended by all team members. The preparatory meeting of the acceptance team should focus on the acceptance requirements, schedule and work division.

6.3.3 The kick-off meeting shall be presided over by the acceptance team leader and attended by all acceptance team members and parties concerned, including project owner, designer, supervisor, and construction contractor. The meeting agenda should include:

    1   Announce the acceptance team members.

    2   State the main basis, procedure and work plan for the acceptance.

    3   Listen to the presentations about the project archives management and

self-inspection results made by the parties concerned, including project owner, supervisor, and construction contractor.

4   Discuss with the parties concerned and inquire about their presentations and self-inspection reports.

6.3.4 After the kick-off meeting, the acceptance team shall pay a site visit to check the physical progress and performance of the project.

6.3.5 The archives acceptance team checks the project archives by means of inquiry, site inspection and spot check. For the spot check of files, emphasis shall be put on the project records regarding initiation, investigation and design, bidding, contracts and agreements, project management, concealed works, quality inspection, defect treatment, supervision, as-built drawings, equipment, acceptance, audios and videos, physical objects, etc. The number of files subjected to spot check shall not be less than 300.

6.3.6 The project owner shall request the parties concerned to designate personnel to support the acceptance team in the archives acceptance.

6.3.7 After the site inspection, the acceptance team shall hold an internal meeting to summarize and discuss the findings and prepare the opinions. The archives acceptance opinions should include:

1   Project overview.

2   Basis for archives acceptance.

3   Project archives management.

4   Problems and suggestions.

5   Acceptance conclusion.

6.3.8 The wrap-up meeting is chaired by the acceptance team leader and attended by all team members and parties concerned, including project owner, designer, supervisor, and construction contractor. The meeting agenda should include:

1   Brief on the archives acceptance activities.

2   Comment on the problems identified.

3   Announce the acceptance opinions.

4   Propose a preliminary rectification plan by the project owner.

6.4 Implementation of Acceptance Opinions

6.4.1 For the project that passes the archives acceptance, the project owner shall rectify the problems identified according to the opinions of the acceptance team.

6.4.2 For the project that fails the archives acceptance, the project owner shall rectify the problems identified within the specified time according to the opinions of the acceptance team. The rectification results shall be submitted to the acceptance organizer for reacceptance.

6.4.3 The project owner shall arrange and file the records generated during the process of archives acceptance.

Appendix A Archives Acceptance Workflow for Hydropower Projects

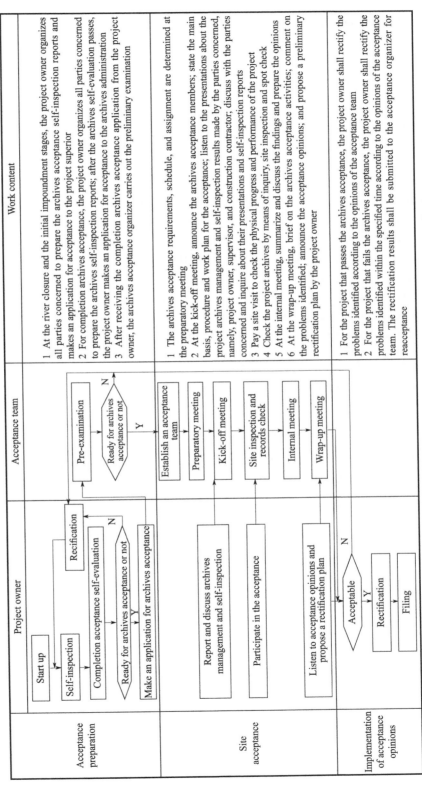

Figure A Archives acceptance workflow for hydropower projects

Appendix B Main Contents of Self-Inspection Report by Project Owner

1 Project Overview

Project location, scale, initiation and approval, administration and management, main parties concerned, lotting scheme for main works, work breakdown, milestones, physical progress, works acceptance, etc.

2 Project Archives Management

Project archives management system, organization and full- or part-time staff, archives management rules and regulations, etc.; measures taken for archives completeness, accuracy, systematicness and security; acceptance scope and the generation, collection, arrangement, filing and transfer of records of project; the compilation of the as-built drawings and its quality; the archives informatization and digitalization; statistics of files on various media by category according to works or contract works; the role of archives in project construction, management, commissioning, etc.

3 Previous Archives Acceptances

Archives acceptance organizers, acceptance conclusions, main problems and their rectifications.

4 Archives Self-Inspection

Self-inspection activities, completion archives self-evaluation, main problems and improvement measures.

5 Overall Evaluation of Project Archives

Completeness, accuracy, systematicness, standardization and security of project archives.

6 Pending Issues

Explain why some records of outstanding works are not yet included in the acceptance.

7 Attachments

　　1) Opinions of previous archives acceptances.

　　2) Self-evaluation form for completion archives acceptance for hydropower project.

Appendix C Main Contents of Self-Inspection Report by Designer

1 Project Overview

Project location, scale, main structures, scope of investigation and design, etc.

2 Delivery of Design Documents

Statistics of design reports, design drawings, and design change notices issued by designer from the start of design to the completion of the project.

3 Design Changes

Important design optimization and major design changes during the project construction.

4 Design Archives Management

Design archives management system, full- or part-time staff on site, archives management rules and regulations; measures taken for archives completeness, accuracy, systematicness and security of the investigation and design archives; acceptance scope and the generation, collection, arrangement, filing and transfer, and quantity of design documents.

5 Main Problems Identified in Previous Archives Acceptances and Self-Inspection and Rectifications

Main problems identified in previous archives acceptances and rectifications; self-inspection activities, main problems found and improvement measures.

6 Overall Evaluation of Design Archives

Completeness, accuracy, systematicness, standardization, and security of design archives.

Appendix D Main Contents of Self-Inspection Report by Supervisor

1   Project Overview

Project location, scale, main structures, supervision scope, etc.

2   Supervision Archives Management

Supervision archives management system, organization and full- or part-time staff, archives management rules and regulations; measures for completeness, accuracy, systematicness and security of supervision archives; acceptance scope and the generation, collection, arrangement, filing and transfer of the supervision records; statistics of transferred archives according to the type and medium of records.

3   Supervision and Guidance on Archival Work and Review of Completion Records

Supervision and guidance on the generation, collection, arrangement and filing of project records; review of the completion archives and as-built drawings, and opinions on completeness, accuracy and systematicness.

4   Main Problems Identified in Previous Archives Acceptances and Rectifications

Main problems identified in previous archives acceptances and rectifications.

5   Completion Archives Self-Inspection, Main Problems and Improvement Measures

Completion archives self-inspection activities, main problems found and improvement measures.

6   Overall Evaluation of Supervision Archives

Completeness, accuracy, systematicness, standardization, and security of supervision archives.

Appendix E Main Contents of Self-Inspection Report by Construction Contractor

1 Project Overview

Project location, scale, main structures, scope and content of construction.

2 Project Archives Management

Project archives management system, organization and full- or part-time staff, archives management rules and regulations; measures for completeness, accuracy, systematicness and security of project archives; acceptance scope and the generation, collection, arrangement, filing, transfer and quantity of project records; the compilation, quality and quantity of as-built drawings transferred; statistics of files on various media by category according to works and contract works.

3 Main Problems Identified in Previous Archives Acceptances and Rectifications

Main problems identified in previous archives acceptances and rectifications.

4 Completion Archives Self-Inspection, Main Problems and Improvement Measures

Completion archives self-inspection activities, main problems found and improvement measures.

5 Overall Evaluation of Construction Archives

Completeness, accuracy, systematicness, standardization, and security of construction archives.

Appendix F   Main Contents of Self-Inspection Report by Units for Testing, Inspection and Safety Monitoring

1   Project Overview

Project location, scale, main structures, contract scope, etc.

2   Project Archives Management

Full- or part-time staff, archives management rules and regulations; measures for completeness, accuracy, systematicness and security; acceptance scope and the generation, collection, arrangement, filing and transfer of the records; statistics of transferred archives according to the type and medium of records.

3   Main Problems Identified in Previous Archives Acceptances and Rectifications

Main problems identified in previous archives acceptances and rectifications.

4   Completion Archives Self-Inspection, Main Problems and Improvement Measures

Completion archives self-inspection activities, main problems found and improvement measures.

5   Overall Evaluation of Archives of Units for Testing, Inspection and Safety Monitoring

Completeness, accuracy, systematicness, standardization, and security of archives of units for testing, inspection and safety monitoring.

Appendix G Main Contents of Archives Self-Inspection Report by EPC Contractor

1   Project Overview

Project location, scale, EPC scope, main subcontractors, milestones, physical progress, etc.

2   Project Management

Project management system, work breakdown, works acceptance, etc.

3   Project Design and Design Changes

Statistics of design reports, design drawings, and design change notices issued by designer from the start of design to the completion of the project; important design optimizations and major design changes during the project construction.

4   Archives Management

Project archives management system, organization and full- or part-time staff, archives management rules and regulations; measures for completeness, accuracy, systematicness and security of the project archives; acceptance scope and the generation, collection, arrangement, filing, transfer and quantity of project records; the compilation, quality and quantity of as-built drawings transferred; statistics of files on various media by category according to the works and contract works.

5   Main Problems Identified in Previous Archives Acceptances and Rectifications

Conclusions and main problems identified in previous archives acceptances and rectifications.

6   Completion Archives Self-Inspection, Main Problems and Improvement Measures

Completion archives self-inspection activities, main problems found and improvement measures.

7   Overall Evaluation of EPC Project Archives

Completeness, accuracy, systematicness, standardization, and security of EPC project archives.

8   Pending Issues

Explain why some records of outstanding works are not yet included in the acceptance.

Appendix H Self-Evaluation Form for Completion Archives Acceptance of Hydropower Projects

Table H Self-evaluation form for completion archives acceptance of hydropower projects

S/N	Acceptance items	Acceptance checklist	Supporting documents	Evaluation criteria	Dominant items	Self-evaluation results	Remarks
1		Archives support system					
1.1	Organization	(1) The person in charge of archival work is designated	Related documents	Rated as "unacceptable" if the requirements are not met	—		
		(2) The archival organization or department is set up and staffed with full- and part-time project archivists	Documents about organization setup, personnel responsibilities and training certificate	Rated as "unacceptable" if no archival organization is set up or no full- or part-time archivists are assigned	√		
		(3) A project archives management network including parties concerned is established and managed by the project owner and the responsible persons are designated	Network charts and documents about implementing the accountability system	Rated as "unacceptable" if the requirements are not met	—		
1.2	Rules and regulations	(1) Rules and regulations for project archives management are established	Relevant regulations and rules	Rated as "unacceptable" if necessary regulations are absent	—		
		(2) The table of filing scope and retention period for project records is developed in line with the actual situation of the project	Table of filing scope and retention period	Rated as "unacceptable" if necessary regulations are absent	—		
		(3) The archive classification scheme is developed in line with the actual situation of the project	Related documents	Rated as "unacceptable" in the case of no archive classification scheme	√		

Table H *(continued)*

S/N	Acceptance items	Acceptance checklist	Supporting documents	Evaluation criteria	Dominant items	Self-evaluation results	Remarks
1.3	Fund	The fund needed for various archival activities of the project owner can meet the demand for archiving	Related supporting documents	Rated as "unacceptable" if the funds do not meet the archiving and archives management needs	—		
1.4	Equipment and facilities	(1) A dedicated archival repository that meets the safekeeping requirement is set up		Rated as "unacceptable" if there is no archival repository	—		
		(2) Equipment/device, facilities/tools can meet the needs of file and archives, and archives retention. The equipment and facilities containers meet the requirements of archiving and security		Rated as "unacceptable" if there are quite a lot of problems with the equipment and facilities	—		
		(3) The office, reading room, and archival repository shall be separated from each other		Rated as "unacceptable" if there is no separation of these three areas	—		
1.5	Concurrency	(1) The collection, arrangement, and transfer of records of project are included in contract management	Related contracts and agreements	Rated as "unacceptable" if not included in the contract management	√		
		(2) The briefing, guidance, training and oversight by the project owner over the supervisor, the construction contractor, etc. concerning document collection and arrangement	Related supporting documents	Rated as "unacceptable" if not providing explanation, guidance, training and supervision and inspection	—		
		(3) The supervision and guidance by the project owner over the departments concerned on document collection, arrangement and filing	Related supporting documents	Rated as "unacceptable" if supervision and guidance not provided	—		

Table H *(continued)*

S/N	Acceptance items	Acceptance checklist	Supporting documents	Evaluation criteria	Dominant items	Self-evaluation results	Remarks
1.5	Concurrency	(4) The project archival work is kept in pace with project construction and included in the quality management procedure; during the completion acceptance of works, section of works and unit of works, the related project documents are inspected or accepted accordingly	Related regulations and records	Rated as "unacceptable" if not concurrent	—		
2		Completeness, accuracy and systematicness of project archives					
2.1		Completeness of project archives					
2.1.1	Completeness of power generation documents	Production preparation documents	Related documents and records	Rated as "unacceptable" if the absence rate is 10 % or above	—		
2.1.2	Completeness of research and development documents	Research and development documents	Relevant reports	Rated as "unacceptable" if the absence rate is 10 % or above	—		
2.1.3	Completeness of project development documents	(1) Preliminary design documents and project initiation documents relating to land requisition, investment and approval documents	Related documents and records	Rated as "unacceptable" if the absence rate is 10 % or above	√		
		(2) Design documents relating to bidding design, construction drawing design and design modification	Investigation and design reports, drawings, design change notices, etc.		—		
		(3) Project management records			—		

Table H *(continued)*

S/N	Acceptance items	Acceptance checklist	Supporting documents	Evaluation criteria	Dominant items	Self-evaluation results	Remarks
2.1.3	Completeness of project development documents	1) Bidding			—		
		2) Contract management			—		
		3) Fund management			—		
		4) Materials management	Related documents and records	Rated as "unacceptable" if the absence rate is 10 % or above	—		
		5) Quality management			√		
		6) Progress, safety, soil and water conservation and environmental protection management			—		
		7) Resettlement			—		
		(4) Supervision records			—		
		1) Construction supervision			—		
		2) Equipment manufacture supervision	Related documents and records	Rated as "unacceptable" if the absence rate is 10 % or above	—		
		3) General resettlement supervision			—		
		4) Environmental protection supervision			—		
		(5) Civil construction records			—		
		1) Water retaining structure	Related documents and records	Rated as "unacceptable" if the absence rate is 10 % or above	√		
		2) Release structure			√		

Table H *(continued)*

S/N	Acceptance items	Acceptance checklist	Supporting documents	Evaluation criteria	Dominant items	Self-evaluation results	Remarks
2.1.3	Completeness of project development documents	3) Water conveyance structure			√		
		4) Powerhouse and switchyard			√		
		5) Transportation, navigation and dam-passing structures	Related documents and records	Rated as "unacceptable" if the absence rate is 10 % or above	—		
		6) Slope works			—		
		7) Ancillary works			—		
		(6) Electromechanical installation records					
		1) Hydraulic turbine-generator units and associated equipment			√		
		2) Hydraulic machinery auxiliaries			—		
		3) Electrical primary equipment	Related documents and records	Rated as "unacceptable" if the absence rate is 10 % or above	√		
		4) Electrical secondary equipment			√		
		5) Utility system			—		
		6) Hydraulic steel structure and hoist			—		
		(7) Commissioning test	Related documents and records	Rated as "unacceptable" if the absence rate is 10 % or above	√		
		(8) Project acceptance	Acceptance certificate		—		

NB/T 10076-2018

Table H *(continued)*

S/N	Acceptance items	Acceptance checklist	Supporting documents	Evaluation criteria	Dominant items	Self-evaluation results	Remarks
2.1.3	Completeness of project development documents	(9) As-built drawing documents	Related drawings and records	Rated as "unacceptable" if the absence rate is 10 % or above; rated as "unacceptable" if description of as-built drawing preparation is absent	—		
		(10) Audio-visual records	Related documents and records	Rated as "unacceptable" if the absence rate is 10 % or above	—		
		(11) Electronic records	Related documents and records	Rated as "unacceptable" if the absence rate is 10 % or above	—		
		(12) Physical archives		Rated as "unacceptable" if the absence rate is 10 % or above	—		
2.1.4	Completeness of equipment and instrument documents	(1) Hydraulic turbine-generator units and associated equipment	Related documents, drawings, and records	Rated as "unacceptable" if the absence rate is 10 % or above	√		
		(2) Hydraulic machinery auxiliaries			—		
		(3) Electrical primary equipment			√		
		(4) Electrical secondary equipment			—		
		(5) Utility system			—		
		(6) Hydraulic steel structure and hoist			—		
		(7) Safety monitoring equipment			—		

Table H (continued)

S/N	Acceptance items	Acceptance checklist	Supporting documents	Evaluation criteria	Dominant items	Self-evaluation results	Remarks
2.2	Accuracy of project archives	(1) The documents generation is standard and the data authentic			—		
		(2) Each type of acceptance evaluation table meets the requirements	Related documents	Rated as "unacceptable" if 30 places or more are found inaccurate	—		
		(3) The different documents about the same issue are consistent in content			—		
		(4) The filed documents shall be original, and the catalog is consistent with the archives			—		
		(5) The as-built drawings are well prepared, reflecting the actual built project clearly and accurately. The as-built drawings are stamped and signed as required and has been reviewed by the supervisor	As-built drawings and records	Rated as "unacceptable" if, on the as-built drawings of the main works, there are 20 or more places that are inconsistent with the actual construction; Rated as "unacceptable" if, on the as-built drawings of other works, there are 30 or more places that are inconsistent with the actual construction; Rated as "unacceptable" if the supervisor has not reviewed the as-built drawings as required	√		
		(6) The documents are legible, the graphs are neat, the review and signing are complete, and the handwriting meets the requirements	Related documents	Rated as "unacceptable" if there are 30 and more places that cannot not meet the relevant requirements	—		

NB/T 10076-2018

Table H *(continued)*

S/N	Acceptance items	Acceptance checklist	Supporting documents	Evaluation criteria	Dominant items	Self-evaluation results	Remarks
2.3	Systematicness of project archives	(1) Reasonable classification. Project archives classification scheme is developed to ensure that the archives are accurately classified and logically arranged	Archives classification scheme, and the file classification	Rated as "unacceptable" if there are 20 or more mistakes	—		
		(2) Reasonable grouping. It follows the principle of subject-oriented grouping and maintains the logical relationship between documents. The file is of the same subject and has appropriate thickness for easy maintenance and use	Grouping	Rated as "unacceptable" if the grouping fails to maintain the inherent logical relation of the generated documents, or there are 20 or more mistakes	—		
		(3) Orderly sequencing. Documents with similar content or closely related are sequenced according to their importance or in chronological order	Innerfile arrangement	Rated as "unacceptable" if the arrangement does not meet the requirement; rated as "unacceptable" if there are 20 or more irregularities	—		
2.4	Archives compilation normalization	(1) The title of file is concise and accurate, the file-item list is up to the standard and the description is accurate	File title and list	Rated as "unacceptable" if there is no file catalog or if there are 30 places or more having problems in cataloging	—		
		(2) The contents and description of the innerfile item list are accurate	Innerfile item list	Rated as "unacceptable" if 10 files or more have no contents or if there are 30 places or more having problems with the innerfile item file list	—		
		(3) The file note is filled in and signed as required	File note	Rated as "unacceptable" if 20 files or more have no file notes or if there are 30 places or more having problem with the file note	—		

Table H *(continued)*

S/N	Acceptance items	Acceptance checklist	Supporting documents	Evaluation criteria	Dominant items	Self-evaluation results	Remarks
2.4	Archives compilation normalization	(4) The drawings meet the folding requirements	Files	Rated as "unacceptable" if 30 places or more fail to meet the requirements	—		
		(5) Files are bound in a secure, tidy and pleasing way	Files	Rated as "unacceptable" if there are 30 places or more having problem with file binding	—		
2.5	Archives transfer	(1) The files transferred are reviewed and go through the formalities as required	Review records	Rated as "unacceptable" if not reviewed; rated as "unacceptable" if there are 20 and more missings in the procedure	—		
		(2) The transfer procedures of the archives are complete	List of transferred files	Rated as "unacceptable" if the transfer procedures are not complete	√		
3	Archive security, accessibility and informatization						
3.1	Archive security	(1) The archival repository meets the requirements		Rated as "unacceptable" if the archival repository does not meet the requirements	—		
		(2) The frames retaining the archives are clearly marked and the arrangement meets the requirements		Rated as "unacceptable" if there are 10 or more unclear marks in the archives frames; rated as "unacceptable" if the arrangement does not meet the relevant specifications	—		

Table H *(continued)*

S/N	Acceptance items	Acceptance checklist	Supporting documents	Evaluation criteria	Dominant items	Self-evaluation results	Remarks
3.1	Archive security	(3) The measures against fire, theft, direct sunlight, water, humidity, insect, dust and high temperature are taken to ensure archive security in the repository; the repository is checked periodically	Work log and repository check records	Rated as "unacceptable" if major potential hazards exist in repository safety management	√		
3.2	Archives accessibility	(1) Retrieval aids have been prepared to facilitate the archives access	Retrieval aids	Rated as "unacceptable" if without retrieval aids	—		
		(2) Archives have been used with use records	Use records	Rated as "unacceptable" if without use records	—		
		(3) The archives borrowing records are complete	Archives borrowing records	Rated as "unacceptable" if without archives borrowing records	—		
3.3	Archives informatization	(1) Archive informatization has been implemented	The performance of archives informatization	Rated as "unacceptable" if archives informatization is not performed	—		
		(2) Archives management software is provided. File-level and document-level catalog database is established. Full-text digitalization of archives has been performed	The use of the software and the database	Rated as "unacceptable" if the archives management software is not provided	—		
		(3) Connection with the LAN of the project owner is realized and the network service is available. Database security measures are taken	Online operation	Rated as "unacceptable" if no network service is provided	—		

Table H (continued)

S/N	Acceptance items	Acceptance checklist	Supporting documents	Evaluation criteria	Dominant items	Self-evaluation results	Remarks

Acceptance criteria: All the dominant items are acceptable, and 80 % or above of the total items are acceptable.

Self-evaluation conclusion: Among the total 18 dominant items, ___ items are unacceptable; among the total 80 items, ___ items are unacceptable; and the acceptable rate is ___.

Conclusion of self-evaluation: ___

The project owner (seal):   Date:

NOTES:

1 The acceptance contents listed in 2.1 of this table shall be in accordance with Appendix B "Project Archives Classification of the Project Owner, the Filing Scope and Storage Unit of the Main Records of Project of the Parties Concerned in DL/T 1396, *Specification of the Documents Collection and Archives Arrangement for Hydropower Construction Projects*.

2 "√" denotes "dominant items".

Appendix J Application Form for Archives Acceptance of Hydropower Projects

Table J Application form for archives acceptance of hydropower projects

Project name			
Project owner			
Project owner's superior			
Project examination and approval organization		Date of approval	
Project budget estimate	CNY	Installed capacity	MW
Number of generating units and unit capacity	____units × ____MW/unit	Total reservoir capacity	m^3
Construction commencement date		Date of putting into operation	
Main designers and construction contractors			
Main supervisors		Turbine-generator unit manufacturer	
Acceptance category	☐ Archives acceptance at the river closure stage ☐ Archives acceptance at the initial impoundment stage ☐ Completion archives acceptance		
Planned archives acceptance date		Planned completion acceptance date	
Contact person		Tel.	
Address/Postcode		Email	
Existing archives (Original)	files	Number of drawing sheets	
Self-inspection opinions of the applicant			(Seal) Date:
Opinions of archives acceptance organizer			(Seal) Date:

Explanation of Wording in This Guide

1   Words used for different degrees of strictness are explained as follows in order to mark the differences in executing the requirements in this guide.

    1) Words denoting a very strict or mandatory requirement:

        "Must" is used for affirmation; "must not" for negation.

    2) Words denoting a strict requirement under normal conditions:

        "Shall" is used for affirmation; "shall not" for negation.

    3) Words denoting a permission of a slight choice or an indication of the most suitable choice when conditions permit:

        "Should" is used for affirmation; "should not" for negation.

    4) "May" is used to express the option available, sometimes with the conditional permit.

2   "Shall meet the requirements of…" or "shall comply with…" is used in this guide to indicate that it is necessary to comply with the requirements stipulated in other relative standards and codes.

List of Quoted Standards

GB/T 11822,	*General Requirements for the File Formation of Scientific and Technological Archives*
NB/T 35083,	*Specification for Drafting As-built Drawing Documents of Hydropower Projects*
DA/T 28,	*Specification for Archival Management of Construction Projects*
DL/T 1396,	*Specification of the Documents Collection and Archives Arrangement for Hydropower Construction Projects*